for Littles

MW01677497

Science Activity Books

MAMMALS

Created by The Good and the Beautiful Team

Cover Illustrated by Sandra Eide
Pages Illustrated by Amanda Gulliver
Design by Phillip Colhouer

goodandbeautiful.com

Lesson 1

Draw a line from each mammal to the food it eats. Color the pictures.

Color the animals in the boxes. Cut out the boxes and tape or paste each animal where it goes in the food chain.

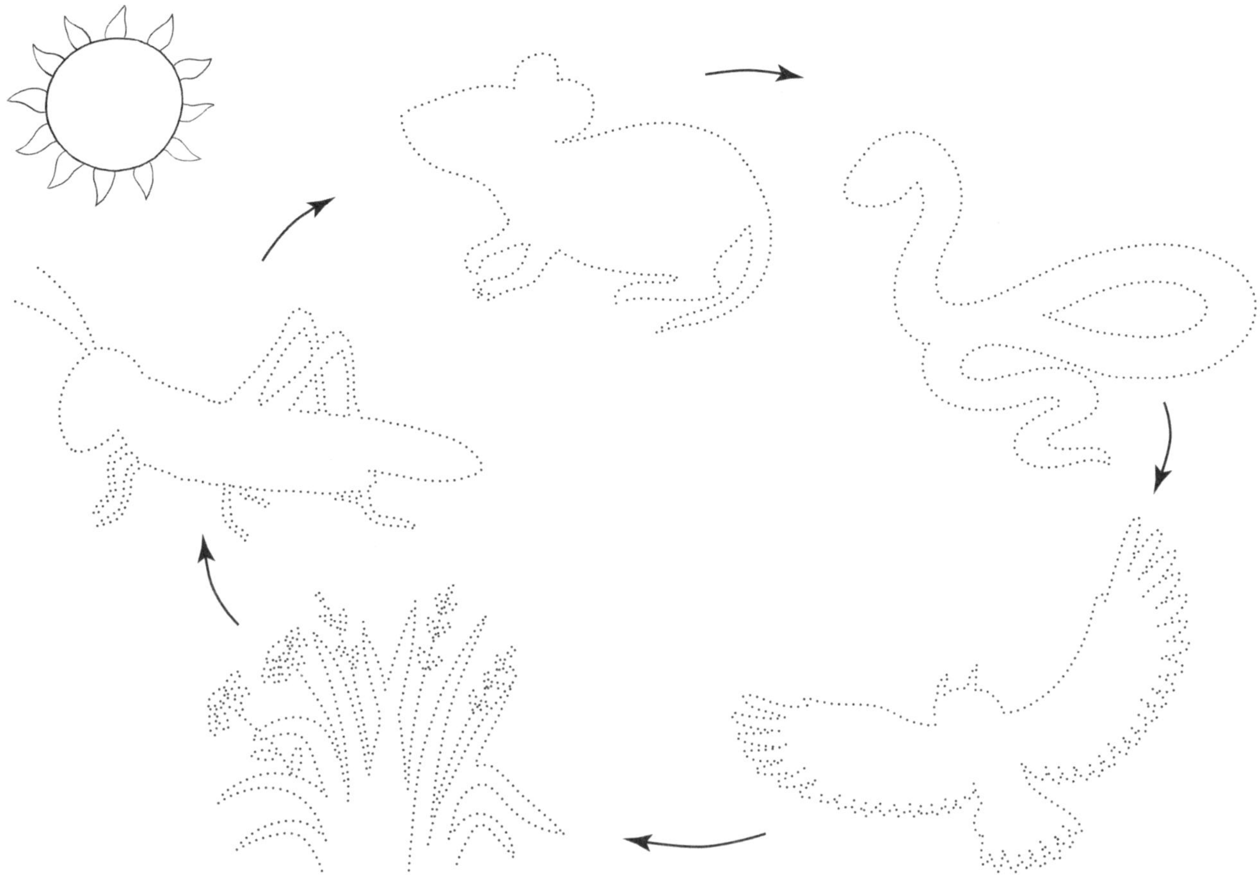

Page intentionally left blank

Page intentionally left blank

Lesson 2

Color the adult and baby mammals. Cut out and tape or paste each baby mammal by its mother.

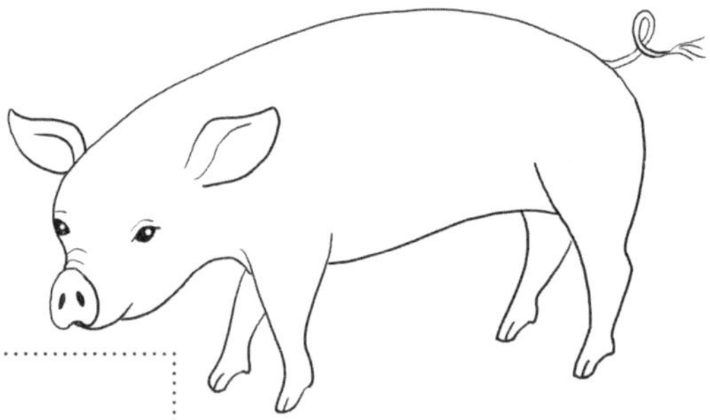

Color the picture of the cat and nursing kittens.

Lesson 3

Trace the dotted lines. Color the picture.

Complete the maze from the horse to the gate. Color the picture.

Lesson 4

For each row, circle and color the animal track that matches the track at the beginning of the row.

Cross out the items in the key as you find them in the forest. Color the forest.

Lesson 5

Connect the dots. Color the elephants.

Find and color the birds and lizards that are in the picture with the elephant. Circle the number of each animal you find in the picture.

Lesson 6

Circle the monkey that matches the first monkey in each row. Color the monkeys.

KEY
Cross out the items in the key as you find them in the jungle. Color the jungle.

Lesson 7

Complete the mazes. Color the animals.

Draw stripes on the large tiger and spots on the large cheetah. Color the pictures. An example tiger and cheetah have been included to show their actual markings.

Lesson 8

Find each animal's tracks in the picture below and place a check in the box. Color the picture.

Cut out the puzzle pieces. Arrange the puzzle correctly. Color the pieces. Tape or paste the puzzle pieces together on the top half of this page.

Page intentionally left blank

Page intentionally left blank

Lesson 9

Cut out the puzzle pieces. Arrange the puzzle correctly. Color the pieces. Tape or paste the puzzle pieces together on the top half of this page.

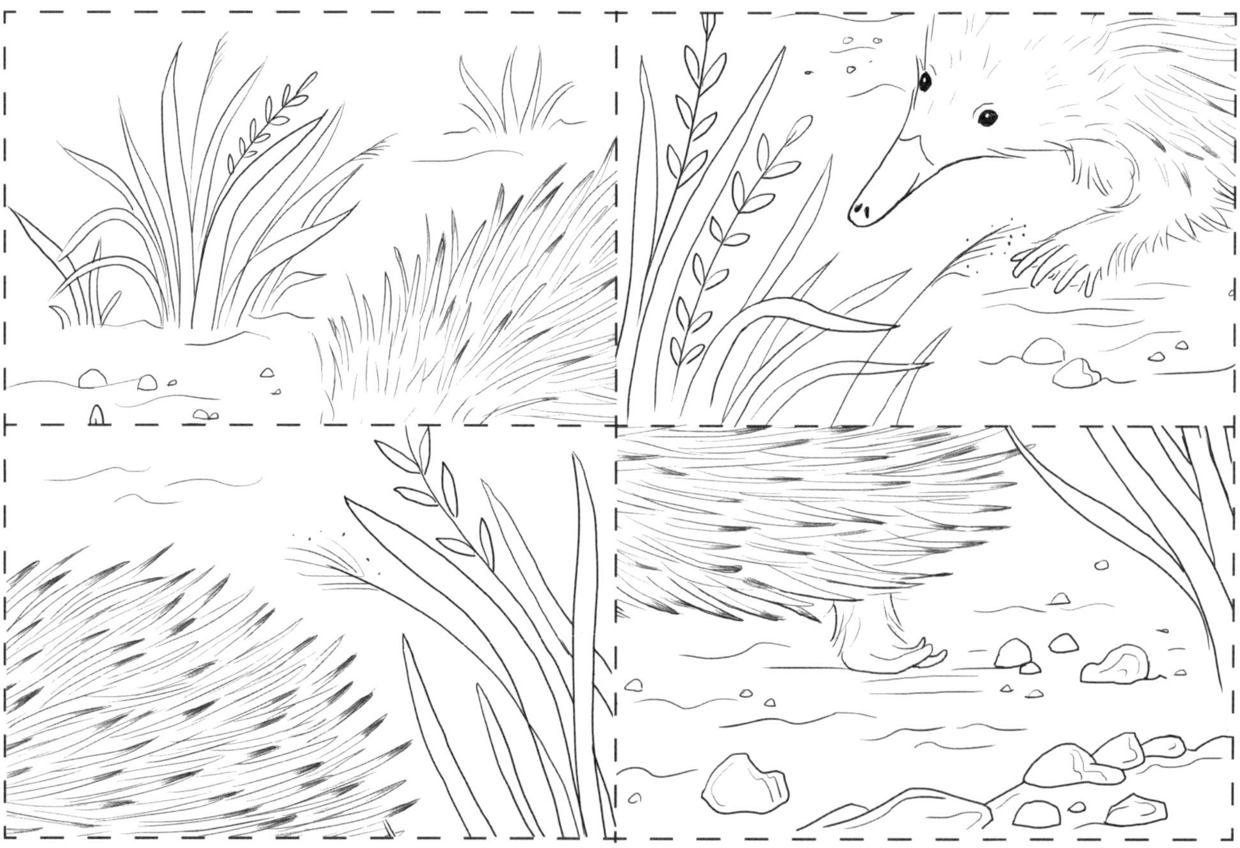

Complete the mazes to help the animals get to their homes.

Lesson 10

Cross out the items in the key as you find them in the forest. Color the forest.

Color the picture and the correct baby that belongs to the kangaroo. Cut out the baby and tape or paste it on the mother's tummy.

Page intentionally left blank

Page intentionally left blank

Lesson 11

Color the mother and baby bats. Cut out the mother bats and tape or paste them under their matching babies.

Connect the dots. Color the bats.

Lesson 12

Circle the four mammals in the top picture that you do not see in the bottom picture. Color the top picture.

Color the number of rodents listed at the beginning of each row.

2

5

3

1

4

Lesson 13

Draw a line connecting each bear to where it lives. Color the bears and their homes.

Cross out the items in the key as you find them in the picture. Color the picture.

Lesson 14

Circle the four animals in the top picture that you do not see in the bottom picture. Color the top picture.

For each row, circle and color the animal track that matches the track at the beginning of each row.

Extra Doodling and Drawing Page

1.ONSP385-155591